o

GRAPHS AND IMAGINARIES

AN EASY METHOD OF FINDING GRAPHICALLY IMAGINARY
ROOTS OF QUADRATIC EQUATIONS AND IMAGINARY
POINTS OF INTERSECTIONS OF VARIOUS CURVES,
WITH ILLUSTRATIONS OF THE PRINCIPLE
FROM ELEMENTARY GEOMETRY

BY

J. G. HAMILTON, B.A.

AND

F. KETTLE, B.A.

AUTHORS OF "A FIRST GEOMETRY BOOK."

LONDON
EDWARD ARNOLD
1904

PREFACE.

IN the following pages we develop a method by which imaginary roots of a quadratic equation can be found graphically as easily and as accurately as real roots, and we extend it to the location of other 'imaginary' points. The principle is clearly capable of much wider application.

As the parabola enables us to find the roots, when real, of any quadratic equation by suitable movement of the axes there would seem some warrant for the conclusion that an intimately related curve would perform a similar function for imaginary roots and thus make the graphical representation complete. Experiment and investigation confirm this inference, and a few typical cases chosen from Elementary and from Analytical Geometry are here worked out in detail in illustration of the principle.

J. G. H.
F. K.

CLAPHAM, S.W.
Jan. 1904.

CONTENTS.

GRAPHS AND IMAGINARIES

I.

Roots of Quadratic Equations from Parabola graph.

Fig. 1 gives the graphs of x^2 and of $-x^2$ and, since the abscissa of any point on the curve is the square root of the ordinate of that point, it is immediately evident from their construction that

$$\sqrt{OK} = \pm\, KM$$

and
$$\sqrt{-OK} = \pm\, i\,.\, KM,$$

kO being equal to OK and km to KM.

In other words, if we move the x-axis up or down, the intercepts between the axis of the parabola and the curve itself give us the roots of the equation $x^2 = n$, for all values of n, positive or negative. Thus we at once read off the approximate roots of

$$x^2 = +3, \text{ as } \pm 1\cdot73,$$

and of
$$x^2 = -3, \text{ as } \pm 1\cdot73\, i,$$

the i being omitted or inserted just as the roots are read off from the real curve (x^2), or its shadow ($-x^2$) shown by the dotted line.

Now since we can express any quadratic equation

$$Ax^2 + Bx + C = 0,$$

in the form $(x+a)^2 = n$, and since also if we have the graph of

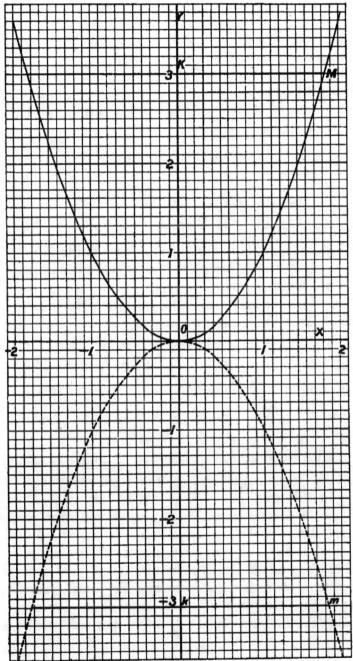

Fig. 1.

x^2 with respect to a pair of axes OX, OY (Fig. 2), we at once get the graph of $(x + a)^2$ by simply moving the y-axis through a distance of a, to the right if a is positive and to left if a is negative, it follows that if we can read off imaginary roots of $x^2 = n$, we can do the same for $(x + a)^2 = n$, i.e. of course for the general equation $Ax^2 + Bx + C = 0$.

The roots of the equation $(x + a)^2 = n$, will be

$O_1O \pm KM$ if n is positive,

$O_1O \pm i . KM$ if n is negative (Fig. 2),

where $O_1O = -a$, and $KM = km = n$.

This is illustrated in Fig. 2 for the case in which $a = -1$, $n = \pm 3$. If OX, OY were the axes the continuous curve would be that of x^2; push the y-axis 1 to the left and regard O_1X, O_1Y_1 as the axes and the continuous curve is now the graph of $(x - 1)^2$ with respect to the new axes. In the figure $O_1O = 1$, $OK = +3$, $Ok = -3$, hence we read off the roots of

$(x - 1)^2 = +3$, or $x^2 - 2x - 2 = 0$, as $1 \pm \sqrt{3}$ or $1 \pm 1 \cdot 73$

and of

$(x - 1)^2 = -3$, or $x^2 - 2x + 4 = 0$, as $1 \pm i . \sqrt{3}$, or $1 \pm 1 \cdot 73i$

approx.

In other words, to find by help of a parabola the roots of an equation given in the form $x^2 - 2x + 4 = 0$, we may plot the graph in the usual way. If we did so, taking O_2X_2 and O_2Y_1 as the axes of x and y, the graph would come in the position shown by the continuous curve in Fig. 2. As this does not cut the x-axis the roots are clearly imaginary, but these can be read off almost as readily as real roots from the intersections of the shadow curve with the x-axis, the roots being $O_2k \pm i . km$. But of course there is no necessity to draw the shadow parabola at all, for we may simply imagine

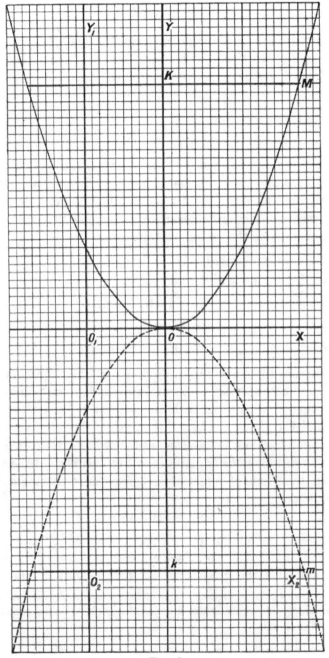

Fig. 2.

the x-axis pushed up through twice the minimum value of the corresponding function*. This latter method is in agreement with the fact that if we take any quadratic equation,

$$ax^2 + bx + c = 0 \quad \dots\dots\dots\dots\dots(1),$$

and form a new equation

$$ax^2 + bx + c = 2m \quad \dots\dots\dots\dots(2),$$

m being the minimum value of $ax^2 + bx + c$, the roots of the derived equation are the same as those of the original except that the discriminant of the derived is i times the discriminant of the original equation. Reverting to the case illustrated by Fig. 2, since twice the minimum value of

$$x^2 - 2x - 2, \text{ i.e. } (x-1)^2 - 3 \text{ is } 2(-3)$$

we have, for original and derived equations,

$$x^2 - 2x - 2 = 0 \quad \dots\dots\dots\dots\dots(1),$$
$$x^2 - 2x - 2 = 2(-3) \quad \dots\dots\dots\dots(2),$$

the roots being $1 \pm \sqrt{3}$, and $1 \pm i\sqrt{3}$ respectively.

Now since the graphic solution of a quadratic equation by the above method resolves itself into finding the points of intersection of a parabola with the axis of x (or with some line parallel to it) we are led on to consider the case in which the straight line has some other direction, and that case in particular in which the straight line lies entirely outside the parabola, the points of intersection being therefore imaginary. The equations

$$y = x - 2, \ y = x^2 - 4x + 5$$

may be taken in illustration. Their graphs, MVN and mn, are given in Fig. 3, the *shadow* parabola (dotted) is also shown, and also the given line mn, first pushed parallel to

---

* The lowest point of the curve being of course graphically determined, if necessary, by the perpendicular bisector of any chord of the parabola parallel to the x-axis.

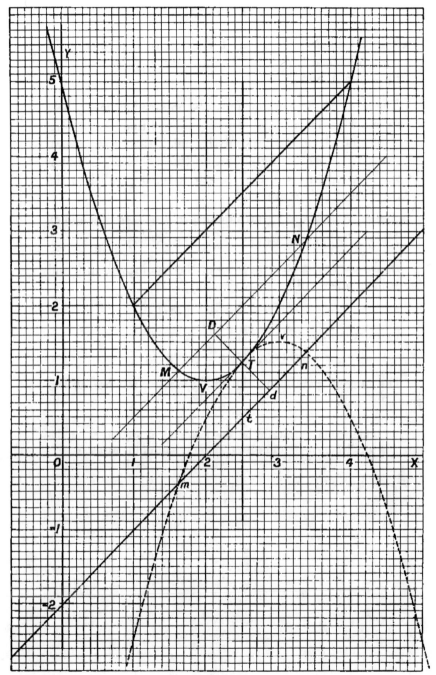

FIG. 3.

itself to a tangential position at T^*, and then further on for an equal distance (so that $TD = dT$), cutting the real curve at the points M and N.

The roots of the equations may be obtained either by help of a shadow parabola which can be obtained from the original by revolution about the tangential point T (and easily shown with tracing paper), or by projecting the points M, N and T on to the given line mn, the projectors being parallel to the y-axis. The readings are :—

For x. Abscissa of $t \pm i$ (difference of abscissae of n and t).

For y. Ordinate of $t \pm i$ („ „ ordinates „ „).

In the particular case illustrated, the readings on inch and tenth-inch paper will be approximately,

$$x = 2{\cdot}5 \pm {\cdot}87i, \quad y = {\cdot}5 \pm {\cdot}87i.$$

By calculation

$$x = 2{\cdot}5 \pm {\cdot}5\sqrt{3}i, \quad y = {\cdot}5 \pm {\cdot}5\sqrt{3}i.$$

The tangent parallel to the line $y = x - 2$, touches at the point $2{\cdot}5$, $1{\cdot}25$.

---

* T being found practically by the bisecting of any chord of the parabola parallel to mn by a line parallel to the axis of the parabola.

II.

Circle method of finding roots of a quadratic equation.

The same principle is readily adapted to what, for brevity, may be called the circle* method of solving quadratic equations. This is illustrated in Fig. 4 for the equation

$$x^2 - 3x + (\tfrac{1}{2})^2 = 0,$$

where $AB = 3$, the sum of the roots,

and $AR \cdot BS = (\tfrac{1}{2})(\tfrac{1}{2})$, the product of the roots,

and in Fig. 5 for the equation $x^2 - 3x + (\tfrac{5}{2})^2 = 0$, AB again representing the sum, and $AR \cdot BS$ the product $(\tfrac{5}{2} \cdot \tfrac{5}{2})$ of the roots.

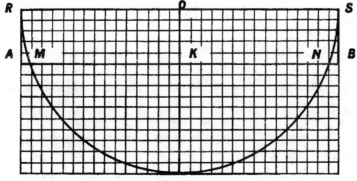

FIG. 4.

* This method of obtaining a first approximation is treated of in H. Bos, *Éléments de Géométrie*, p. 162 (3rd edition, 1888); Chrystal, *Introduction to Algebra*, p. 332; Lodge, *Differential Calculus for Beginners*, p. 91, &c.; Mackay's *Euclid*, VI, 28, 29.

If the circle cuts AB in M and N we have *real* roots which can be read $AK \pm KN$; if it does *not* cut AB the roots are *imaginary*. Taking $x^2 - ax + b = 0$ as our standard equation (a and b positive), keeping a constant, and giving b continually increasing values, we shall find

(1) that the circle gets further and further away from AB,

(2) that the roots are

$AK \pm$ a continually decreasing quantity,

then $AK \pm 0$ (when the circle touches AB),

next $AK \pm i$ times a continually increasing quantity,

and it will be readily perceived that the '*shadow*' circle that is to give us the roots cannot be the mere reflection of the other, but one that will accommodate itself to its *real* but retiring companion. Thus it will be found that for such an equation as $x^2 - 3x + 5^2 = 0$, the radius of the shadow circle is 5, and its centre $\frac{3}{2}$ below AB, and generally, for the equation

$$x^2 - ax + \left(\frac{a}{2} + k\right)^2 = 0,$$

the radius of the shadow circle is $\dfrac{a}{2} + k$, and the centre of the shadow circle is $\dfrac{a}{2}$ below AB, when AR and BS are made equal.

If, however, we adopt the more general and serviceable method of Fig. 6, taking $x^2 - 3x + 6 = 0$ as the equation, and making $AR = 2$, $BS = 3$, we at once see that the statement is not general enough. The modification needed is that if the radius of the real circle is r, the centre of the shadow circle must be taken at a distance of r on the other side of AB, so that for the general equation $x^2 - px + mn = 0$, the radius of the shadow circle is $\frac{1}{2}(m + n)$, and the centre of the shadow circle is r below AB, r, the radius of the real circle, being equal to

$$\sqrt{\frac{p^2}{4} + \frac{(m-n)^2}{4}}.$$

This includes the special case considered above, and in every case the general reading of imaginary roots by the circle

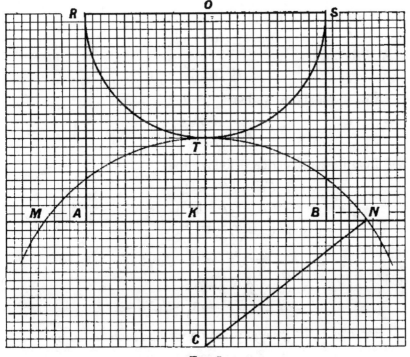

Fig. 5.

method is $AK \pm i.KN$. The *shadow* circle is observed to meet the needs of its *real* comrade by assuming for its radius a length equal to the radius of its companion + the distance which the tangential point of the real circle moves away from AB, and by cutting AB in points M and N which recede

2—2

farther and farther from K as the real circle retires, thus giving the increasing multiple of i in the roots demanded by the increase in the absolute term mn. The general proof of the correctness of the method of reading off the roots for these two cases is appended.

For Fig. 5, the general equation may be represented by

$$x^2 - ax + b = 0,$$

and since the roots are

$$a/2 \pm \sqrt{a^2/4 - b},$$

$i \cdot KN$ should be

$$\sqrt{a^2/4 - b}.$$

Now in the figure we have

$$AB = a,$$
$$AR = BS = \sqrt{b},$$

and from its construction it follows that

$$CK = TO = a/2,$$
$$CN = CT = KO = \sqrt{b},$$
$$\therefore i \cdot KN = i\sqrt{CN^2 - CK^2}$$
$$= i\sqrt{b - a^2/4} = \sqrt{a^2/4 - b}$$

and

$$AK = a/2,$$
$$\therefore AK \pm i \cdot KN = a/2 \pm \sqrt{a^2/4 - b}.$$

For Fig. 6 the general equation may be represented by

$$x^2 - px + mn = 0,$$

and the roots being

$$p/2 \pm \sqrt{p^2/4 - mn},$$

$i \cdot KN$ should be $\sqrt{p^2/4 - mn}.$

In the figure, $AB = p$, $AR = n$, $BS = m$, and, drawing a perpendicular from R to KO, it is seen that radius of original circle $= \sqrt{(KO - AR)^2 + AK^2}$

$$= \sqrt{(m - n)^2/4 + p^2/4} = CK,$$

and $\qquad CN = CT = KO = \tfrac{1}{2}(m + n),$

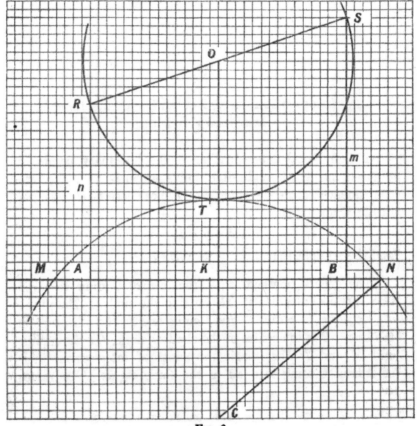

Fig. 6.

$$\therefore i \cdot KN = i \sqrt{CN^2 - CK^2}$$

$$= i \sqrt{(m + n)^2/4 - [(m - n)^2/4 + p^2/4]}$$

$$= i \cdot \sqrt{mn - p^2/4} = \sqrt{p^2/4 - mn},$$

and $\qquad AK = p/2,$

$$\therefore AK \pm i \cdot KN = p/2 \pm \sqrt{p^2/4 - mn}.$$

III.

Intersection of a line and a circle in imaginary points.

The graphical determination of the imaginary points of intersection of a straight line and a circle is an immediate consequence of the foregoing. A typical case is represented in Fig. 7 for the circle

$$x^2 + y^2 = 1,$$

and the straight line $\qquad y = \tfrac{3}{4}x + 3.$

The general method is :—From centre O draw to given line AB a perpendicular OK cutting the circle at T, and set off KC equal to radius of given circle and on the other side of the line. C is the centre of the shadow circle, CT the radius, and M and N the imaginary points of intersection, their co-ordinates being read off in the form $a \pm ib$. The values are

For x. Abscissa of $K \pm i$ (difference of abscissae of K and N).

For y. Ordinate of $K \pm i$ („   ordinates   „   ).

Graph reading on inch paper gives

$$x = -1\cdot44 \pm 1\cdot74\,i,$$
$$y = +1\cdot94 \pm 1\cdot29\,i,$$

which correspond nearly to the calculated results,

$$x = -1\cdot44 \pm 1\cdot746i \atop y = +1\cdot92 \pm 1\cdot309i \Big\}\ .$$

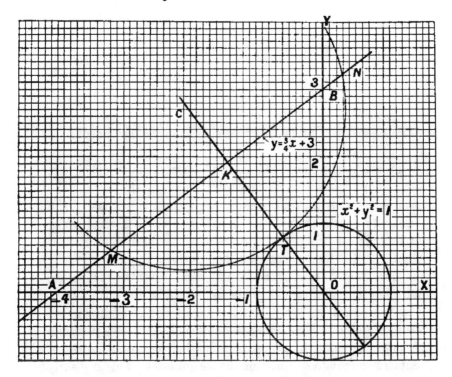

$y = \frac{3}{4}x + 3$

$x^2 + y^2 = 1$

FIG. 7.

The radius, co-ordinates of the centre, and the equation of the shadow circle are readily found by calculation, and its x-points of intersection with the straight line determined as

$$- 1\cdot44 \pm \cdot16 \sqrt{119},$$

the x-points of intersection of the *real* circle and the straight line being $- 1\cdot44 \pm \cdot16 \sqrt{119i}.$

IV.

Intersection of two circles in imaginary points.

Proceeding on somewhat similar lines with two equal circles, keeping one fixed with its centre at the origin and moving the other away with its centre along the x-axis, the locus of the imaginary points of intersection is found to be a rectangular hyperbola; on bringing the movable circle back into contact again with the fixed circle and then pushing it on to the left, the locus is at first, of course, the fixed circle for the real values and then the other branch of the hyperbola for the imaginaries. This is interesting but does not suggest the construction we are in search of. Now, since the common chord always passes through the points of intersection for the 'real' case, its representative ought to do duty for the imaginaries. This is clearly the radical axis, and the radical axis meets all the demands made on it. Making it serve as the AB line in the circle method for roots of quadratics, and getting shadow circles in a similar way for each of the real ones, it is found that the intersections of *either* shadow circle with the radical axis give the required points of intersection. This is illustrated by Fig. 8, the circles there represented being

$$\left. \begin{array}{l} (x-1)^2 + (y-1)^2 = 1^2 \quad A \\ (x-4)^2 + (y-4)^2 = 2^2 \quad B \end{array} \right\}$$

MN is the radical axis and K its point of intersection with the centre line O_1O_2. C_1, the centre of A's shadow circle, is found by setting off KC_1 equal to radius of A. Similarly for

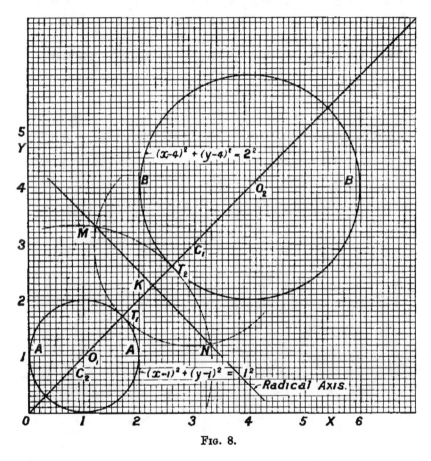

FIG. 8.

C_2. C_1T_1 and C_2T_2 are the radii of the shadow circles. Readings much as before.

For x. Abscissa of $K \pm i$ (difference of abscissae of N and K).

For y. Ordinate of $K \pm i$ (,, ordinates ,,).

By calculation $x = 2{\cdot}25 \pm 1{\cdot}031i,$

$$y = 2{\cdot}25 \mp 1{\cdot}031i.$$

These statements are easily verified by finding the equations of the shadow circles and of the radical axis and obtaining their points of intersection by Algebra. In the case of *each* shadow circle, the equations giving the required points reduce to

$$\left. \begin{array}{c} x + y = \tfrac{9}{2} \\ x^2 - \tfrac{9}{2}x + 4 = 0 \end{array} \right\} .$$

V.

Tangent to a circle from a point within and its imaginary points of contact.

A method of procedure is again suggested here by the principle of continuity. Starting with the usual method of drawing a tangent to a circle with centre O from an external point P we construct a circle on OP as diameter, and the intersections of this circle with the original give us the two points of contact. Imagine P to move up to the circle along a centre line. The tangential points gradually approach one another and coincide at the moment when P reaches the circle. Now if P continues its course towards the centre of the circle it is reasonable to suppose, from our previous work, that the tangential points should begin to recede from one another, probably at a greater rate, along a perpetually accommodating companion circle, and just as, in the common case of a tangent from an external point, the points of intersection of the real circles give us the tangential points, so we expect our two shadow circles to do us a like service when the given point is internal and the points of intersection imaginary. So they do and they appear always to intersect on the radical axis of the two real circles. But now a disturbing element comes in. Just as the join of the external P with a tangential point touches the real circle, we may reasonably expect that when P becomes an internal point its join with an imaginary tangential point will touch the shadow of the original circle.

Compasses and eyes give sufficient testimony on this score, and calculation confirms by proving that *both* the shadow circles are *cut* by the line we have been imagining to be a tangent to one of them. Further investigation establishes the results summarised below. It will be convenient however to precede this summary with a description of the figure (Fig. 9). A is the given circle with centre O, and P the internal point from which tangents to A are to be drawn; B is the circle with OP as diameter [A is taken with a radius of 3, its equation being $x^2 + y^2 = 9$, and P is here taken $2\frac{1}{4}$ from O; OP makes an angle of 45° with the x-axis], MN is the radical axis of A and B. The shadow circles of A and B are obtained in the usual way, KC_2 being equal to A's radius, and KC_1 equal to radius of B, M and N are the points of intersection of the shadow circles with the radical axis, and K the point where the centre line through P cuts the radical axis.

(1) The radical axis of the circles A and B is the polar of P with respect to the original circle A*, and, however purely imaginary the imaginary points of tangency are, Coordinate Geometry requires the polar to pass through them. The equation of the polar is $x + y = 4\sqrt{2}$, the tangential points are $2\sqrt{2} \pm \frac{1}{2}\sqrt{14}i$, $2\sqrt{2} \mp \sqrt{14}i$ and, dropping the i, these are the points given by the intersections, with the polar MN, of both the shadow circles.

(2) If the radius of the original circle A is a, and

   „   „   „ inner „ B „ b,

---

* For, if radii of A and B (Fig. 9) are a and b respectively, O_1 the centre of B, and KM the polar of P with respect to A, then

$$\because OP \cdot OK = a^2, \quad \therefore OK = a^2/2b, \quad \therefore OK^2 - O_1K^2 = (a^2/2b)^2 - (a^2/2b - b)^2$$
$$= b(a^2/b - b) = a^2 - b^2$$

     = difference of squares on radii of A and B,

$\therefore KM$ is also the radical axis.

29

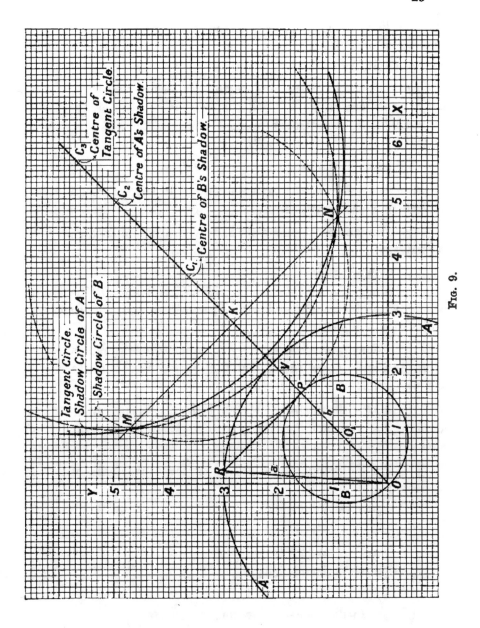

Tangent Circle.

Shadow Circle of A.

Shadow Circle of B.

C_3 Centre of Tangent Circle.

C_2 Centre of A's Shadow.

C_1 Centre of B's Shadow.

Fɪɢ. 9.

and the distance of the polar from the centre of A is d, then the radius of A's shadow circle is d, and the radius of B's shadow circle is $d - b$.

(3) The centre of the circle actually touched, which, for our present purpose, we will call the *tangent circle*, is always twice as far from centre of original circle as the polar is, so that $$KC_3 = OK = d,$$ and the radius of the tangent circle is $\sqrt{2d^2 - a^2}$.

It is easily proved that B's shadow circle passes through the centre of the tangent circle.

(4) The tangential points M and N can be obtained in several ways :—

I.   Along the polar of P set off $KM = KR$.

II.   Along the centre line set off KC_2 equal to a, the radius of original circle, and with C_2 as centre and C_2V as radius obtain A's shadow circle.

III.   Along centre line set off KC_1 equal to b, the radius of B, and with C_1 as centre and C_1P as radius obtain B's shadow circle.

(5) KM is the inverse of the circle B with respect to A and also the inverse of B's shadow with respect to the tangent circle.

(6) The following results are given for convenience of reference :—

$$\begin{cases} C_1 \text{ centre of } B\text{'s shadow,} \\ C_2 \quad \text{,,} \quad \text{,, } A\text{'s} \quad \text{,,} \\ C_3 \quad \text{,,} \quad \text{,, tangent circle.} \end{cases}$$

$$\begin{cases} KC_1 = b = \text{radius of circle } B, \\ KC_2 = a = \quad \text{,, } \quad \text{,,} \quad \text{,, } A, \\ KC_3 = d = \text{distance of polar from } A\text{'s centre.} \end{cases}$$

$$\begin{cases} PC_1 = d - b = (2d^2 - a^2)/2d = \text{radius of } B\text{'s shadow,} \\ VC_2 = OK = KC_3 = d \quad = \quad \text{,,} \quad \text{,, } A\text{'s ,,} \\ MC_3 = \sqrt{2d^2 - a^2} \qquad = \quad \text{,,} \quad \text{,, tangent circle.} \end{cases}$$

$$KM = \sqrt{d^2 - a^2}.$$

$$PM = \sqrt{(d^2 - a^2)(2d^2 - a^2)/d^2} \text{ (actual length).}$$

$$VK = d - a.$$

$$PK = d - 2b = (d^2 - a^2)/d.$$

$$PC_3 = (d^2 - a^2)/d + d = (2d^2 - a^2)/d.$$

The relation between the various circles is seen by considering how they change as P moves from V to O, the circle A being fixed.

In all positions KM is always equal to KR, and KC_3 to OK, and the rect. $OP.OK$ is constant, being always equal to a^2.

When P is at V, then R, K and M are also at V, and the two shadow and the tangent circles have a minimum value, C_2 and C_3 coinciding at a distance of a from K and their circles being each equal to the original circle A, while C_1 is at its maximum distance from K although its circle has a minimum area, its radius being half the radius of the original circle A.

As P moves towards the centre O, R also moves to the left along the circle A; K (and with it M) and the three centres C_1, C_2, C_3 move to the right at various rates, C_1 getting nearer and nearer to K, C_2 and C_3 further and further away. All this while C_2 moves exactly at the same rate as K, the distance between them always being a, the radius of the original circle, but the rate of C_3 is *double* that of K, the distances of K from V being represented by 0, 1, 2, 3 ...... n, and the corresponding distances of C_3 from V by

$$1, 3, 5, 7 \ldots\ldots (2n-1),$$

the radius of the original circle being unity.

Suppose K to have moved on till it is 1000 units from O (radius of A, 1 unit),

A's shadow circle, centre C_2, will have a radius of 1000 exactly.

B's shadow circle, centre C_1, will have a radius of

$$(1000 - \tfrac{1}{2} \cdot \tfrac{1}{1000}).$$

Tangent circle, centre C_3, will have a radius of

$$1000\sqrt{2}\sqrt{1 - \tfrac{1}{2} \cdot \tfrac{1}{1000^2}}.$$

VI.

Relation between a real circle and its growing shadow.

(1) P and K (Fig. 10) are points in the centre line of circle A, radius a, such that $OP \cdot OK = a^2$, $KC = OR = a$, and

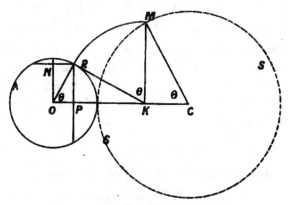

Fig. 10.

circle S with centre C touches A at the point V. S is therefore the shadow circle of A.

From the construction, however obtained, the following conclusions are readily deduced :—

(a) P and K are inverse points with respect to the circle A.

(b) The triangle $KCM \equiv$ the triangle ROK.

H.   3

(c)   The triangle KCM is similar to the triangle POR.

(d)   The angles marked θ are equal.

Now if we suppose a point to start from V and to move either towards or away from O, the centre of the real circle A, the following relation always holds for the real circle and its shadow. Denoting by d the distance at any moment of the point P (or of K if the movement is outwards) from the centre O, and speaking of PR (or KM) as the perpendicular, then *for the real case* (the one always considered in Elementary Geometry) the perpendicular is $\sqrt{a^2 - d^2}$, a real quantity, and *for the imaginary case* (that in which the geometrical construction is usually said to break down) the perpendicular is $\sqrt{d^2 - a^2}$, and since $i\sqrt{d^2 - a^2} = \sqrt{a^2 - d^2}$, it follows that if we let the same algebraical formula, *perp.* $= \sqrt{a^2 - d^2}$, which is true for the special case, hold generally, the shadow circle invariably gives the geometrical solution, $i.KM$ corresponding strictly to the algebraical result. This is illustrated later by one or two geometrical problems.

(2)   *For a real circle A, radius a,*

     the radius of the shadow circle = $a \sec \theta$,

for $CM = KC \sec \theta = a \sec \theta.$

Consider the changes in the triangles POR, KCM as P moves from V to centre O.

In every position $KC = a = OR$.

When P is at V, then K, R and M are also at V and the angles marked θ are $0°$.

As P moves towards O, K moves outwards,

    θ increases from $0°$ to $90°$, and $\sec \theta$ from 1 to ∞,

    $\therefore$ radius of shadow circle increases from a to ∞.

(3) *The parts of the polars of P and K* intercepted by the real circle and its shadow are as the radii of the circles,*

i.e. as $\qquad a : a \sec \theta$, or as $1 : \sec \theta$.

$$\therefore KM = PR \sec \theta.$$

Limiting cases : P *at* V. $\quad KM = PR \sec \theta = 0\,(1) = 0.$

$\qquad\qquad\quad P$ *at* O. $\quad KM = a \sec \theta \qquad = \infty.$

(4) *The locus of M is a hyperbola,* P moving from V to v, and M denoting a point of intersection of A's shadow circle with the polar of P (Fig. 11); vV is the diameter of A.

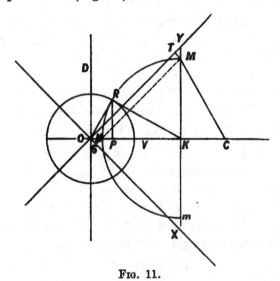

FIG. 11.

Taking as axes OX, OY at angles of 45° with OK, it is easy to show that the rect. $OS \cdot SM$ is constant for all positions of M. For, if the circle MRN is described with K as centre, the triangle NKM is obviously isosceles.

Also the triangle OKY is isosceles, angle at O being 45°.

* Here and throughout, always with reference to the original circle A, unless otherwise stated.

On producing MN to S, it is seen that the angle S is 90°, and the triangle OSN equal to the right-angled triangle MTY.

Also the rect. $OS.SM = ONMY$

$$= \text{triangle } OKY - \text{triangle } NKM$$
$$= \tfrac{1}{2}OK^2 - \tfrac{1}{2}NK^2$$
$$= \tfrac{1}{2}(OK^2 - RK^2)$$
$$= \tfrac{1}{2}OR^2$$
$$= \tfrac{1}{2}a^2.$$

The equation of the curve is therefore $xy = \tfrac{1}{2}a^2$, a hyperbola referred to the asymptotes OX, OY as axes, and with V as vertex. When P passes the centre O, KM appears on the left and we get the other branch of the hyperbola with its vertex at v. Referred to OC, OD as axes, the equation of the hyperbola is $x^2 - y^2 = a^2$, the equation of the circle A being $x^2 + y^2 = a^2$, also referred to OC, OD as axes.

If, in the light of this, we now reconsider the question raised on p. 24 as to the points of intersection of two equal circles, one with its centre fixed at the origin, equation $x^2 + y^2 = a^2$, the other moving right and left with its centre always on the x-axis, we shall see that the locus of the points of intersection is always given by the equation

$$x^2 \pm (iy)^2 = a^2.$$

VII.

Application to Elementary Geometry.

1.   *Divide* 10 *into two parts so that their product shall be* 40.

This problem is given by Cardano (1545) and is interesting as being apparently the earliest on record to start the quest after imaginaries (see Chrystal's *Algebra*, Part I. p. 248, Historical Note).

A geometrical solution, answering to the algebraical, can be obtained by help of shadow circle thus :—

Construct Fig. 6, making AB, AR, BS, 10, 5, 8 half-inches respectively, and read off from shadow circle as on p. 19. The calculated result is $5 \pm \sqrt{15}i$, or $5 \pm 3\cdot873i$ approx.

2.   A similar example and fairly typical of the algebraic geometrical problems suggested by Euc. II. 11 and 14, is

Divide a line 4 inches long into two parts so that the sum of the squares on them shall be 4 sq. ins.

Denoting the parts by x and $4 - x$ we get

$$x^2 - (4 - x)^2 = 4,$$

i.e. $$x^2 - 4x + 6 = 0,$$

which is worked as in Fig. 6, the roots being read as

$$2 \pm 1{\cdot}41i.$$

3. *Given a circle, radius a, find length of chord which is distant d from the centre,* (i) *when* $d < a$, (ii) *when* $d > a$.

Denoting the length of half the chord by z, then for the ordinary case in which the distance d from the centre is less than the radius a, we have $z = \sqrt{a^2 - d^2}$, and the corresponding geometrical construction is to set off $OP = d$ along a centre line OC (Fig. 10) and to draw the perpendicular PR to meet the circle, in which case $PR = z = \sqrt{a^2 - d^2}$.

If we now inquire what is the geometrical significance of the question when the lengths have been so chosen that $d > a$, the reply is that although we can find no chord *within* the given circle to satisfy the conditions, any more than we can, by keeping to the primitive meaning of the terms, 'divide a straight line 5 cm. long into two parts whose difference is 7 cm.,' yet just as in the latter case a geometrical construction can be found in harmony with the algebraical statement of the case, so in the case under present consideration can geometry, by means of the shadow circle, be made to harmonise with its fleeter companion Algebra.

The construction is just as in the 'real' case, OK, Fig. 10, being set off equal to d, and the length of the perpendicular KM that meets the shadow circle, when multiplied by i answers fully to the algebraical result for

$$KM = \sqrt{CM^2 - KC^2}$$
$$= \sqrt{d^2 - a^2},$$
$$\therefore i \cdot KM = \sqrt{a^2 - d^2} = z, \text{ just as before.}$$

4. *Given a circle, radius a, find how far distant from the centre is a chord 2c units in length, (i) when c < a, (ii) when c > a.*

Worked much as last example; Fig. 10 illustrates the method. For the ordinary case, twice RN is the chord of length $2c$, ON or PR is the required distance.

For the 'imaginary' case, set off as before a length equal to c; the point P (now shown as K in the figure) falls outside the given circle, therefore have recourse to the shadow circle and $i \cdot KM$ gives the required length.

5. *Euc.* II. *5, 6 and 14 and the introduction of the imaginary.*

When Euc. II. 5 and 6 are looked at with modern eyes accustomed to negatives and imaginaries, and account taken of the 'sense' of the lines, then instead of being regarded as two distinct propositions their identity is seen to be closer even than geometry books have often indicated when giving their algebraical representation. This essential likeness would suggest that when we have a geometrical construction, as in II. 14, based on the more primitive conception of II. 5 and giving us a positive root of $\sqrt{(+a)(+b)}$, then the introduction of the negative idea, besides making the old construction yield us the negative root, would give completion by leading us on in due course to the imaginary. The corresponding representation of the imaginary roots of $\sqrt{(+a)(-b)}$ is the inevitable consequence, and readily follows with the aid of II. 6, the negative twin brother of II. 5, and of the shadow circle with its power of adapting itself to all conceivable changes in a and b. Not only will the real identity of these two propositions be observed in what follows, but the construction for imaginary roots will be seen to follow closely in the steps of that for real roots, justified too with an almost identical proof.

Fig. 12 suggests the usual construction and proof of Euc. II. 14 by means of II. 5, but for comparison with the companion problem for imaginary roots the following observations may be made :—

$$DP = a, \quad PE = b \text{ (Fig. 12)}.$$

The diameter DE of the circle with centre $O = a + b$,

$$\therefore \text{ the radius } OR = \tfrac{1}{2}(a+b),$$
$$\text{and } OP = \tfrac{1}{2}(a-b).$$

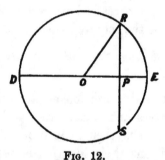

Fig. 12.

The perpendicular from which the roots are read is drawn from P, the common extremity of a and b, PR giving the positive and PS the negative root.

The actual length of $PR = \sqrt{OR^2 - OP^2}$

$$= \sqrt{\tfrac{1}{4}(a+b)^2 - \tfrac{1}{4}(a-b)^2}$$
$$= \sqrt{ab} = \sqrt{(+a)(+b)}.$$

Fig. 13 shows the construction for the 'imaginary' case.

Using the same letters as before, $DP = a, \ PE = -b$.

The diam. DE of the circle with centre $O = a + (-b)$,

$$\therefore \text{ the radius } OR_1 = \tfrac{1}{2}[a + (-b)] = \tfrac{1}{2}(a-b),$$
$$\text{and } OP = \tfrac{1}{2}[a - (-b)] = \tfrac{1}{2}(a+b).$$

The perp. from which the roots are read is drawn from P, the common extremity of a and b, PR as before giving the positive and PK the negative root. Here, however, P is outside the real circle, centre O, so the shadow circle is called

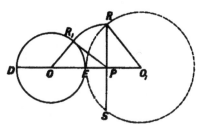

into requisition, its centre O_1 being determined by setting off, along the centre line, PO_1 equal to radius of real circle, its radius being O_1E.

It is at once evident that

$$O_1R = EO_1 = OP = \tfrac{1}{2}(a+b)$$

and $$PO_1 = OR_1 = \tfrac{1}{2}(a-b).$$

Hence the actual length of $PR = \sqrt{O_1R^2 - PO_1^2}$
$$= \sqrt{\tfrac{1}{4}(a+b)^2 - \tfrac{1}{4}(a-b)^2},$$
$$\therefore\; i \,.\, PR = \sqrt{\tfrac{1}{4}(a-b)^2 - \tfrac{1}{4}(a+b)^2}$$
$$= \sqrt{(+a)(-b)}.$$

In short, for the 'imaginary' case the only difference is that in the algebraical part $-b$ is substituted for $+b$, while in the geometrical construction the perpendicular is drawn to meet the shadow instead of the real circle with the consequent reading of $\pm i \,.\, PR$ for the result*.

* To complete this it may be added that an apt illustration of the geometrical significance of $\sqrt{(-a)(-b)}$ is readily given by a piece of tracing paper. If Fig. 12 (segments DP, PE being $+a$, $+b$) is marked through on tracing paper and then swung about the point P through an angle of $180°$ the effect of the swing is to negative both a and b, and the lines representing the roots are seen to coincide with those of $\sqrt{(+a)(+b)}$ so that $\sqrt{(-a)(-b)} = \sqrt{(+a)(+b)}$.

𝕮𝖆𝖒𝖇𝖗𝖎𝖉𝖌𝖊:

PRINTED BY J. AND C. F. CLAY,

AT THE UNIVERSITY PRESS.

The Elements of Geometry

By R. LACHLAN, Sc.D.,

Formerly Fellow of Trinity College, Cambridge,

AND

W. C. FLETCHER, M.A.,

Chief Inspector of Secondary Schools.

With about 750 Exercises and Answers. Clearly printed and furnished with a large number of bold and distinct Diagrams.

Crown 8vo. pp. xii + 207. 2s. 6d.

This book has been prepared in accordance with the recommendations of the Mathematical Committee of the British Association, and on the lines laid down in the Syllabus lately issued by Cambridge University for the Local Examinations.

Elementary Geometry

Containing the whole substance of EUCLID I.-IV., IV., except the elegant but unimportant Proposition IV. 10.

By W. C. FLETCHER, M.A.,

Chief Inspector of Secondary Schools, late Fellow of St John's College, Cambridge.

84 pages. Crown 8vo. Cloth. Price 1s. 6d.

The Mathematical Gazette.—"We cannot speak too highly of Mr Fletcher's work. It has been constructed with admirable skill, and every teacher will find it most suggestive."

Nature.—"This is a very small book and a very good one."

Plane Geometry

Adapted to Heuristic Methods of Teaching.

By T. PETCH, B.A., B.Sc.,

Lecturer in Mathematics, Leyton Technical Institute.

viii + 112 pages. Cloth, 1s. 6d.

LONDON : EDWARD ARNOLD.

ARNOLD'S SCHOOL SERIES.

An entirely new Text-book of Arithmetic, based upon the
most recent Methods of Instruction.

An Arithmetic for Schools

By J. P. KIRKMAN, M.A. CAMBRIDGE,

AND

A. E. FIELD, M.A. OXFORD,

Assistant Masters at Bedford Grammar School.

500 pages, Crown 8vo., with or without Answers, cloth, 3s. 6d.

The work is issued in Two Editions, WITH ANSWERS (RED COVER)
and WITHOUT ANSWERS (GREEN COVER) at a uniform price.
The Edition with Answers will be supplied if no definite in-
structions are given.

In this volume the object has been to provide a thoroughly up-to-
date Standard Text-book of Arithmetic, proceeding logically
through the subject, and treating it in such a manner as to
promote good habits of thinking and a due understanding of the
processes which underlie the practical use of numbers.

Educational Times.—" An excellent school arithmetic on a good plan
well carried out. In addition to its value as a text-book, the work is
attractive from the quantity of instructive general information it con-
tains. This is so simply and directly given that a student can scarcely
fail to be thoroughly interested and vividly impressed. The theory of
the elementary rules is exceptionally clear and convincing."

Nature.—" The work is one that can be very confidently recommended
to all teachers and students of arithmetic."

EXERCISES IN ARITHMETIC (Oral and Written)
PARTS I. AND II.

By C. M. TAYLOR (Mathematical Tripos, Cambridge), Wimbledon
High School. With or without Answers, cloth, 1s. 6d. each.
Answers alone, 6d.

The special features of this work are the absence of long me-
chanical exercises, and the use of decimals at a very early stage,
before vulgar fractions. (*Part III. in preparation.*)

LONDON: EDWARD ARNOLD.

ARNOLD'S SCHOOL SERIES.

ENGLISH.

EPOCHS OF ENGLISH LITERATURE. By J. C. Stobart, M.A., Assistant Master at Merchant Taylors' School, formerly Scholar of Trinity College, Cambridge. In eight volumes. Price 1s. 6d. each.

> Vol. I. **The Chaucer Epoch, 1215-1500.**
> Vol. II. **The Spenser Epoch, 1500-1600.**
> Vol. III. **The Shakespeare Epoch, 1600-1625.**
> Vol. IV. **The Milton Epoch, 1625-1660.**

[Others in preparation.

ARNOLD'S SCHOOL SHAKESPEARE. Issued under the General Editorship of Professor J. Churton Collins.

Price 1s. 3d. each.

Macbeth.	Julius Cæsar.	The Merchant of
Twelfth Night.	Midsummer Night's	Venice.
As You Like It.	Dream.	The Tempest.

Price 1s. 6d. each.

King Lear.	Richard III.	Hamlet.
Richard II.	King John.	
Henry V.	Coriolanus.	

ARNOLD'S BRITISH CLASSICS FOR SCHOOLS. Issued under the General Editorship of Professor Churton Collins.

Paradise Lost, Books I. and II. 1s. 3d.	The Lay of the Last Minstrel. 1s. 3d.
Paradise Lost, Books III. and IV. 1s. 3d.	The Lady of the Lake. 1s. 6d.
	Childe Harold. 2s.
Marmion. 1s. 6d.	Macaulay's Lays of Ancient Rome. 1s. 6d.

ARNOLD'S ENGLISH TEXTS. A series of texts of English Classics to which nothing has been added but a small glossary of archaic or unusual words. Paper, 6d. each ; cloth, 8d. each.

I. MACBETH.	III. THE TEMPEST.	V. TWELFTH NIGHT.
II. HENRY V.	IV. AS YOU LIKE IT.	VI. CORIOLANUS.

STEPS TO LITERATURE. A Graduated Series of Reading Books for Preparatory Schools and Lower Form Pupils. Seven books, prices 10d. to 1s. 6d. With beautiful Illustrations, many of them being reproductions of Old Masters.

IN GOLDEN REALMS. An English Reading Book for Junior Forms. 224 pages. Illustrated. Crown 8vo., cloth, 1s. 3d.

IN THE WORLD OF BOOKS. An English Reading Book for Middle Forms. 256 pages. Illustrated. Crown 8vo., cloth, 1s. 6d.

THE GREENWOOD TREE. A Book of Nature Myths and Verses. 224 pages. Illustrated, 1s. 3d.

LAUREATA. Edited by Richard Wilson, B.A. Crown 8vo., cloth, 1s. 6d. Beautifully printed and tastefully bound.
A collection of gems from the best poets from Shakespeare to Swinburne.

TELLERS OF TALES. Edited by Richard Wilson, B.A. Crown 8vo., cloth, 1s. 6d.
Biographies of some English novelists, with Extracts from their works.

POETS' CORNER. Selected Verses for young Children. Fcap. 8vo., 1s.

LONDON : EDWARD ARNOLD, 41 & 43 MADDOX STREET, W.

ARNOLD'S SCHOOL SERIES.

ENGLISH.

Selections from Matthew Arnold's Poems. Edited with Introduction and Notes, by RICHARD WILSON, B.A. Cloth, 1s. 6d.

Selections from the Poems of Tennyson. Edited, with Introduction and Notes, by the Rev. E. C. EVERARD OWEN, M.A. 1s. 6d.

A Book of Poetry. Edited by M. B. WARRE CORNISH. 200 pages. Cloth, 1s. 3d. Also in Three Parts, paper covers, 4d. each.

Lingua Materna. By RICHARD WILSON, B.A. 3s. 6d.

A First Course in English Literature. By RICHARD WILSON, B.A. 144 pages. Crown 8vo., 1s.

A First Course in English Analysis and Grammar. By RICHARD WILSON, B.A. 144 pages. Crown 8vo., 1s.

Exercises for Parsing in Colour. By EDITH HASTINGS, Headmistress of Wimbledon High School. Cloth, 1s. 6d. Also in Three Parts, each containing Colour Chart, paper covers, 6d. each.

HISTORY.

A History of England. By C. W. OMAN, M.A., Chichele Professor of Modern History in the University of Oxford. Fully furnished with Maps, Plans of the Principal Battlefields, and Genealogical Tables. 760 pages. Thirteenth Edition (to end of South African War). Crown 8vo., cloth, 5s.
    Special Editions, each volume containing a separate index.
    In Two Parts, 3s. each : Part I., from the Earliest Times to 1603 ; Part II., from 1603 to 1902.
    In Three Divisions : Division I., to 1307, 2s. ; Division II., 1307 to 1688, 2s. ; Division III., 1688 to 1902, 2s. 6d.
    ** *In ordering please state the period required, to avoid confusion.*

England in the Nineteenth Century. By C. W. OMAN, M.A., Author of "A History of England," etc. With Maps and Appendices. Revised and Enlarged Edition. crown 8vo., 3s. 6d.

A Junior History of England. From the Earliest Times to the Death of Queen Victoria. By C. W. OMAN, M.A., and MARY OMAN. With Maps. Cloth, 2s.

Questions on Oman's History of England. By R. H. BOOKEY, M.A. Crown 8vo., cloth, 1s.

A Synopsis of English History. By C. H. EASTWOOD, Headmaster of Redheugh Board School, Gateshead. 2s.

Seven Roman Statesmen. A detailed Study of the Gracchi, Cato, Marius, Sulla, Pompey, Cæsar. Illustrated with reproductions of Roman Coins from the British Museum. By C. W. OMAN. About 320 pages. Crown 8vo., cloth, 6s.

English History for Boys and Girls. By E. S. SYMES, Author of "The Story of Lancashire," "The Story of London," etc. With numerous Illustrations. 2s. 6d.

LONDON : EDWARD ARNOLD, 41 & 43 MADDOX STREET, W.

HISTORY.

Men and Movements in European History. Illustrated.
Small crown 8vo., 1s. 6d.

Britain as Part of Europe. 256 pages. Illustrated. 1s. 6d.

Wardens of Empire. 224 pages. Fully illustrated. 1s. 6d.

The Pageant of the Empires. 256 pages. Fully illustrated. 1s. 6d.

Lessons in Old Testament History. By the Venerable
A. S. AGLEN, Archdeacon of St. Andrews, formerly Assistant Master at
Marlborough College. 450 pages. with Maps. Crown 8vo., cloth, 4s. 6d.

Old Testament History. By the Rev. T. C. FRY, Headmaster of Berkhamsted School. Crown 8vo., cloth, 2s. 6d.

GEOGRAPHY.

Arnold's Home and Abroad Atlas.
Containing 24 full-page (11½ × 9 inches) maps, printed in colour. Bound
in a stout paper wrapper. Price 8d. net.

CONTENTS.

1. The World.	9. Spain and Portugal.	18. Africa.
2. Europe.	10. Italy.	19. North America.
3. The United Kingdom.	11. Scandinavia.	20. South America.
4. Scotland.	12 & 13. England & Wales.	21. Canada.
5. Ireland.	14. The Balkan Peninsula	22. The United States.
6. France.	15. European Russia.	23. Australia.
7. The German Empire.	16. Asia.	24. South Africa, New
8. Austria-Hungary.	17. India.	Zealand, Tasmania.

The London School Atlas.
Edited by the Right Hon. H. O. ARNOLD-FORSTER, M.P. A magnificent
Atlas, including 48 pages of Coloured Maps. The size of the Atlas
is about 12 by 9 inches, and it is issued in the following editions:

Stout paper wrapper, with cloth strip at back, 1s. 6d.	Cloth cut flush, 2s. 6d.
	Limp cloth, 3s.
Paper boards, 2s.	Cloth gilt, bevelled edges, 3s. 6d.

A Manual of Physiography.
By ANDREW HERBERTSON, Ph.D., F.R.G.S., Reader in Regional
Geography at the University of Oxford. Fully Illustrated. 4s. 6d.

Arnold's New Shilling Geography.
The World, with special reference to the British Empire.

The World's Great Powers—Present and Past.
Britain, France, Germany, Austria-Hungary, Italy, Russia, The
United States, and Japan. Beautifully Illustrated. 1s. 6d.

The World's Trade and Traders.
The World's Divisions—Some Great Trading Spheres—Some Great
Trading Centres—Some World Routes. Beautifully Illustrated. 1s. 6d.

Arnold's Geographical Handbooks.
A Series of 10 little Manuals providing accurate and clearly-arranged
summaries of Geographical information. 3d. each; cloth, 6d.

LONDON: EDWARD ARNOLD, 41 & 43 MADDOX STREET, W.

ARNOLD'S SCHOOL SERIES.

MATHEMATICS AND SCIENCE.

Arnold's Shilling Arithmetic. By J. P. KIRKMAN, M.A., and J. T. LITTLE, M.A., Assistant Masters at Bedford Grammar School. Crown 8vo., cloth, 1s.

A New Arithmetic for Schools. By J. P. KIRKMAN, M.A., and A. E. FIELD, M.A., Assistant Masters at Bedford Grammar School. cloth, 3s. 6d.

Exercises in Arithmetic (Oral and Written). Parts I., II., and III. By C. M. TAYLOR (Mathematical Tripos, Cambridge), Wimbledon High School. Cloth, 1s. 6d. each. (With or without Answers.)

Mensuration. By R. W. K. EDWARDS, M.A., Lecturer on Mathematics at King's College, London. Cloth, 3s. 6d.

The Elements of Algebra. By R. LACHLAN, Sc.D. With or without Answers, 2s. 6d. Answers separately, 1s.

Algebra for Beginners. By J. K. WILKINS, B.A., and W. HOLLINGSWORTH, B.A. In Three Parts. Part I., 4d. ; Part II., 4d. ; Part III., 6d. Answers to Parts I.-III., in one vol., 6d.

The Elements of Geometry. By R. LACHLAN, Sc.D., and W. C. FLETCHER, M.A. With about 750 Exercises and Answers. Cloth, 2s. 6d.

Elementary Geometry. By W. C. FLETCHER, M.A., Crown 8vo., cloth, 1s. 6d.

A First Geometry Book. By J. G. HAMILTON, B.A., and F. KETTLE, B.A. Crown 8vo., fully illustrated, cloth, 1s.

A Second Geometry Book. By J. G. HAMILTON and F. KETTLE. [In the press.

Elementary Solid Geometry. By F. S. CAREY, M.A., Professor of Mathematics in the University of Liverpool. 2s. 6d.

Geometrical Conics. By G. W. CAUNT, M.A., Lecturer in Mathematics, Armstrong College, Newcastle-on-Tyne, and C. M. JESSOP, M.A., Professor of Mathematics, Armstrong College, Newcastle-on-Tyne. Crown 8vo., 2s. 6d.

The Elements of Euclid. By R. LACHLAN, Sc.D.

Book I. 145 pages, 1s.	Books I.—IV. 346 pages, 3s.
Books I. and II. 180 pages, 1s. 6d.	Books III. and IV. 164 pages, 2s.
Books I.—III. 304 pages, 2s. 6d.	Books I.—VI. and XI. 500 pages. 4s. 6d.
Books IV.—VI. 2s. 6d.	Book XI. 1s.

Test Papers in Elementary Mathematics. By A. CLEMENT JONES, M.A., Ph.D., and C. H. BLOMFIELD, M.A., B.Sc., Mathematical Masters at Bradford Grammar School. 250 pages. Crown 8vo., without Answers, cloth, 2s. 6d. ; with Answers, 3s. Answers separately, 1s.

Vectors and Rotors. With Applications. By Professor O. HENRICI, F.R.S. Edited by G. C. TURNER, Goldsmith Institute. Crown 8vo., 4s. 6d.

A Note-Book of Experimental Mathematics. By C. GODFREY, M.A., Headmaster of the Royal Naval College, Osborne, and G. M. BELL, B.A., Senior Mathematical Master, Winchester College. Fcap. 4to., paper boards, 2s.

An Elementary Treatise on Practical Mathematics. By JOHN GRAHAM, B.A. Crown 8vo., cloth, 3s. 6d.

Preliminary Practical Mathematics. By S. G. STARLING, A.R.C.Sc., and F. C. CLARKE, A.R.C.Sc., B.Sc. 1s. 6d.

An Introduction to Elementary Statics (treated Graphically). By R. NETTELL, M.A., Assistant Master, Royal Naval College, Osborne Fcap. 4to., paper boards, 2s.

Graphs and Imaginaries. By J. G. HAMILTON, B.A., and F. KETTLE, B.A. Cloth, 1s. 6d.

The Principles of Mechanism. By H. A. GARRATT, A.M.I.C.E., Crown 8vo., cloth, 3s. 6d.

The Elements of Trigonometry. By R. LACHLAN, Sc.D., and W. C. FLETCHER, M.A. Crown 8vo., viii+164 pages, 2s.

Mathematical Drawing. By Professor G. M. MINCHIN, Cooper's Hill Engineering College, and J. B. DALE, Assistant Professor of Mathematics at King's College, London.

LONDON : EDWARD ARNOLD, 41 & 43 MADDOX STREET, W.

ARNOLD'S SCHOOL SERIES.

Mechanics. A Course for Schools. By W. D. EGGAR, Science Master, Eton College. Crown 8vo., 3s. 6d.

Electricity and Magnetism. By C. E. ASHFORD, M.A., Headmaster of the Royal Naval College, Dartmouth, late Senior Science Master at Harrow School. With over 200 Diagrams. Cloth, 3s. 6d.

Magnetism and Electricity. By J. PALEY YORKE, of the Northern Polytechnic Institute, Holloway. Crown 8vo., cloth, 3s. 6d.

A Preliminary Course of Practical Physics. By C. E. ASHFORD, M.A., Headmaster of the Royal Naval College, Dartmouth. Fcap. 4to., 1s. 6d.

Advanced Examples in Physics. By A. O. ALLEN, B.A., B.Sc., A.R.C.Sc. Lond., Assistant Lecturer in Physics at Leeds University. 1s. 6d.

A Text-Book of Physics. By Dr. R. A. LEHFELDT. Cloth, 6s.

The Elements of Inorganic Chemistry. For use in Schools and Colleges. By W. A. SHENSTONE, F.R.S., Lecturer in Chemistry at Clifton College. New Edition, revised and enlarged. 554 pages. Cloth, 4s. 6d.

A Course of Practical Chemistry. By W. A. SHENSTONE. Cloth, 1s. 6d.

A First Year's Course of Experimental Work in Chemistry. By E. H. COOK, D.Sc., F.I.C., Principal of the Clifton Laboratory, Bristol. Crown 8vo., cloth, 1s. 6d.

A Text-Book of Physical Chemistry. By Dr. R. A. LEHFELDT. With 40 Illustrations. Crown 8vo., cloth, 7s. 6d.

Physical Chemistry for Beginners. By Dr. VAN DEVENTER. Translated by Dr. R. A. LEHFELDT. 2s. 6d.

The Standard Course of Elementary Chemistry. By E. J. Cox, F.C.S., Headmaster of the Technical School, Birmingham. In Five Parts, issued separately, bound in cloth and illustrated. Parts I.-IV., 7d. each; Part V., 1s. The complete work in one vol., 3s.

First Steps in Quantitative Analysis. By J. C. GREGORY, B.Sc., A.I.C. Crown 8vo., cloth, 2s. 6d.

Oblique and Isometric Projection. By J. WATSON. Fcap. 4to., 3s. 6d.

Physiology for Beginners. By LEONARD HILL, M.B. 1s.

A Text-Book of Zoology. By G. P. MUDGE, A.R.C.Sc. Lond., Lecturer on Biology at the London Hospital Medical College. Illustrated. Crown 8vo., 7s. 6d.

A Class-Book of Botany. By G. P. MUDGE, A.R.C.Sc., and A. J. MASLEN, F.L.S. With Illustrations. Crown 8vo., 7s. 6d.

Psychology for Teachers. By C. LLOYD MORGAN, F.R.S., Principal of University College, Bristol. New, Revised and Enlarged Edition. Crown 8vo., 4s. 6d.

The Laws of Health. By DAVID NABARRO, M.D., B.Sc., Assistant Professor of Pathology and Morbid Anatomy at University College, London. Crown 8vo., 1s. 6d.

LONDON : EDWARD ARNOLD, 41 & 43 MADDOX STREET, W.

ARNOLD'S SCHOOL SERIES.

GERMAN.

DER BACKFISCHKASTEN. By FEDOR VON ZOBELTITZ. Edited, with Notes and Vocabulary, by GUSTAV HEIN, German Master at the High School for Girls, Aberdeen, N.B. Authorised Edition. Crown 8vo., cloth, 2s.

EASY GERMAN TEXTS. For pupils who have acquired a simple vocabulary and the elements of German. Under the General Editorship of WALTER RIPPMANN, M.A., Professor of German at Queen's College, London. With exercises on the text. Small crown 8vo., cloth, 1s. 3d. each.

    ANDERSEN'S BILDERBUCH OHNE BILDER (What the Moon Saw).
    PRINZESSIN ILSE. By MARIE PETERSEN.
    DER TOPFER VON KANDERN. By H. VILLINGER.
    DIE FLUT DES LEBENS. By ADOLF STERN.

A FIRST GERMAN READER. With Questions for Conversation, Grammatical Exercises, Vocabulary, &c. Edited by D. L. SAVORY, B.A., Lecturer in the University of London, Goldsmiths' College. Crown 8vo., cloth, 1s. 6d.

GERMAN WITHOUT TEARS. By Lady BELL. A version in German of "French Without Tears." With illustrations. Cloth.
    Part I., 9d.   Part II., 1s.   Part III., 1s. 3d.

LESSONS IN GERMAN. A graduated German Course, with Exercises and Vocabulary, by L. INNES LUMSDEN, late Warden of University Hall, St. Andrews. Crown 8vo., 3s.

KLEINES HAUSTHEATER. Fifteen little Plays in German for Children. By Lady BELL. Crown 8vo., cloth, 2s.

FRENCH.

ARNOLD'S MODERN FRENCH BOOK I. Edited by H. L. HUTTON, M.A., Senior Modern Languages Master at Merchant Taylors' School. Crown 8vo., cloth, 1s. 6d.

ELEMENTS OF FRENCH COMPOSITION. By J. HOME CAMERON, M.A., Lecturer in French in University College, Toronto, Canada. viii+196 pages. Crown 8vo., cloth, 2s. 6d.

LE FRANCAIS CHEZ LUI. A French Reader on Reform Lines, with Exercises on Grammar for Middle and Junior Forms. By W. H. HODGES, M.A., Modern Language Master at Merchant Taylors' School, and P. POWELL, M.A., Assistant Master at Loretto School. Cloth, 1s. 3d.

MORCEAUX CHOISIS. French Prose Extracts. Selected and Edited by R. L. A. DU PONTET, M.A., Assistant Master in Winchester College. Explanatory Notes and Short Accounts of the Authors cited are given. Crown 8vo., cloth, 1s. 6d.

POEMES CHOISIS. Selected and Edited by R. L. A. DU PONTET, M.A. Cloth, 1s. 6d.

LES FRANCAIS EN MÉNAGE. By JETTA S. WOLFF. With Illustrations. 1s. 6d. An entirely original book, teaching the ordinary conversation of family life in France by a series of entertaining scenes.

LES FRANCAIS EN VOYAGE. By JETTA S. WOLFF. Cleverly illustrated. Crown 8vo., cloth, 1s. 6d.

FRANCAIS POUR LES TOUT PETITS. By JETTA S. WOLFF. With Illustrations by W. FOSTER. Cloth, 1s. 3d.

LES FRANCAIS D'AUTREFOIS. Stories and Sketches from the History of France. By JETTA S. WOLFF. Cloth, 1s. 3d.

LES FRANCAIS DU DIX-HUITIÈME SIÈCLE. By JETTA S. WOLFF. With Notes and Vocabulary. Cloth, 1s. 3d.

LONDON : EDWARD ARNOLD, 41 & 43 MADDOX STREET, W.

ARNOLD'S SCHOOL SERIES.

GRAMMAIRE FRANCAISE. A l'usage des Anglais. Par E. RENAULT, Officier d'Académie; Assistant Lecturer at the University of Liverpool. viii+360 pages. Crown 8vo., cloth, 4s. 6d.

FRENCH WITHOUT TEARS. A graduated Series of French Reading Books, carefully arranged to suit the requirements of quite young children beginning French. With Humorous Illustrations, Notes, and Vocabulary. By Lady BELL. Book I.. 9d ; Book II.. 1s. ; Book III.. 1s. 8d

GRADUATED FRENCH UNSEENS. Edited by Professor VICTOR OGER, Professor in French at Bedford College for Women, London. In four parts. Limp cloth, 8d. each.

A FIRST FRENCH COURSE. Complete, with Grammar, Exercises and Vocabulary. By JAMES BOÏELLE, B.A. (Univ. Gall.). Cloth, 1s. 6d.

A FIRST FRENCH READER. With Exercises for Re-translation. Edited by W. J. GREENSTREET, M.A., Head Master of the Marling School, Stroud. Crown 8vo., cloth, 1s.

FRENCH DRAMATIC SCENES. By C. ABEL MUSGRAVE. With Notes and Vocabulary. Crown 8vo., cloth, 2s.

ARNOLD'S FRENCH TEXTS. An entirely new series of texts, graduated in difficulty, with notes and vocabulary. General Editor: MAURICE A. GEROTHWOHL, B.Litt., L.-ès-L., F.R.L.S., Examiner to the Central Welsh Board. Limp cloth, 6d. each.

Le Forçat ou A tout Péché Miséricorde. Prover's in two acts. By MADAME DE SÉGUR. 48 pages.

Aventures de Tom Pouce. By P. J. STAHL. 48 pages.

L'Histoire de la Mère Michel et de son Chat. By COMTE E. DE LA BEDOLLIÈRE. 48 pages.

Gribouille. By GEORGES SAND. 48 pages.

Laurette ou Le Cachet rouge. By ALFRED DE VIGNY. 48 pages.

La Souris blanche et Les Petits Souliers. By HÉGÉSIPPE MOREAU. 48 pages.

La Vie et ses de Polichinelle et ses Nombreuses Aventures. By OCTAVE FEUILLET. 48 pages.

Crispin rival de son Maître. Comedy in one act. By LE SAGE. 48 pages.

Le Bon Père. Comedy in one act. By FLORIAN. 64 pages.

Monsieur Tringle. By CHAMPFLEURY. 48 pages.

Aventures du Chevalier de Grammont. By Chevalier D'HAMILTON. 48 pages.

Histoire d'un Pointer écossais. By ALEXANDRE DUMAS père. 48 pages.

Deux Heroines de la Revolution. Madame Roland and Charlotte Corday. By JULES MICHELET. 48 pages.

Trafalgar. By JOSEPH MÉRY. 48 pages.

Marie Antoinette. By EDMOND and JULES DE GONCOURT. 48 pages.

Mercadet. A Comedy in three acts. By H. DE BALZAC. 64 pages.

SIMPLE FRENCH STORIES. An entirely new series of easy texts, with Notes, Vocabulary, and Table of Irregular Verbs, prepared under the General Editorship of Mr. L. VON GLEHN, Assistant Master at Perse School, Cambridge. About 80 pages in each volume. Limp cloth, 9d.

Un Drame dans les Airs. By JULES VERNE.

Pif-Paf. By EDOUARD LABOULAYE.

La Petite Souris Grise; and Histoire de Rosette. By MADAME DE SÉGUR.

Poucinet, and two other tales. By EDOUARD LABOULAYE.

Un Anniversaire à Londres, and two other stories. By P. J. STAHL.

Monsieur le Vent et Madame la Pluie. By PAUL DE MUSSET.

La Fée Grignotte. By Madame DE GIRARDIN. And La Cuisine au Salon. From Le Théâtre de Jeunesse.

Gil Blas in the Den of Thieves. Arranged from LE SAGE. With Notes and Vocabulary by R. DE BLANCHAUD, B.A., Assistant Master at the Central Schools, Aberdeen. Limp cloth, crown 8vo., 9d. [*Uniform with the above series.*

L'APPRENTI. By EMILE SOUVESTRE. Edited by C. F. HERDENER, French Master at Berkhamsted School. Crown 8vo., cloth, 1s.

RICHARD WHITTINGTON. By MADAME EUGENIE FOA. And UN CONTE DE L'ABBE DE SAINT-PIERRE. By EMILE SOUVESTRE. Edited by C. F. HERDENER. Crown 8vo., cloth, 1s.

MEMOIRES D'UN ANE. By MADAME DE SÉGUR, edited by Miss LUCY E. FARRER, Assistant in French at the Bedford College for Women, London. Cloth. Crown 8vo., 1s.

LONDON : EDWARD ARNOLD, 41 & 43 MADDOX STREET, W.

ARNOLD'S SCHOOL SERIES.

LATIN.

ARNOLD'S LATIN TEXTS. General Editor, A. EVAN BERNAYS, M.A. The object of the series is to supply short texts, adapted for lower forms, sufficient to provide one term's work. Each volume consists of a short introduction, text and vocabulary. 64 pages. Cloth limp, 8d. each.

HORACE. — Odes, Book I. By L. D. WAINWRIGHT, M.A., Assistant Master at St. Paul's School.

OVID. — **Selections.** By GEORGE YELD, M.A.

OVID IN EXILE.—Selections from the 'Tristia.' By L. D. WAINWRIGHT, M.A.

CORNELIUS NEPOS. — Select Lives. By L. D. WAINWRIGHT, M.A.

VERGIL.—Select Eclogues. By J. C. STOBART, M.A., Assistant Master at Merchant Taylors' School.

VERGIL. — Selections from the Georgics. By J. C. STOBART, M.A.

PHÆDRUS.—Select Fables. By Mrs. BROCK, formerly Assistant Mistress at the Ladies' College, Cheltenham.

TIBULLUS.—Selections. By F. J. DOBSON, B.A., Lecturer at Birmingham University.

CÆSAR in BRITAIN. By F. J. DOBSON, B.A.

CICERO.—In Catilinam, I. and II. By L. D. WAINWRIGHT, M.A.

CICERO.—Pro Archia. By Mrs. BROCK.

LIVY. — **Selections.** By R. M. HENRY, M.A., Classical Master at the Royal Academical Institution, Belfast.

VIRGIL—ÆNEID. Books I., II., and III. The New Oxford Text, by special permission of the University. Edited, with Introduction and Notes, by M. T. TATHAM, M.A. Crown 8vo., cloth, 1s. 6d. each.

CÆSAR'S GALLIC WAR. Books I. and II. Edited by T. W. HADDON, M.A., and G. C HARRISON, M.A. With Notes, Maps, Plans, Illustrations, Helps for Composition, and Vocabulary. Cloth, 1s. 6d.
Books III.-V. Edited for the use of Schools by M. T. TATHAM, M.A. Uniform with Books I. and II. Crown 8vo., cloth, 1s. 6d.
Books VI. and VII. By M. T. TATHAM, M.A. Uniform with Books III.-V. 1s 6d.

LIVY. Book XXVI. Edited, with Introduction and Notes, by R. M. HENRY, M.A. Cloth, 2s. 6d.

THE FABLES OF ORBILIUS. By A. D. GODLEY, M.A., Fellow of Magdalen College, Oxford. With humorous Illustrations. Crown 8vo., cloth. Book I., 9d.; Book II., 1s.

EASY LATIN PROSE. By W. H. SPRAGGE, M.A., Assistant Master at the City of London School. Cloth, 1s. 6d.

DIES ROMANI. A new Latin Reading Book. Edited by W. F. WITTON, M.A., Classical Master at St. Olave's Grammar School. Cloth, 1s. 6d.

SENTENCES FOR LATIN COMPOSITION. Based upon the Exercises in "Fables of Orbilius, Part II." By Rev. A. JAMSON SMITH, M.A., Headmaster of King Edward's Grammar School, Birmingham. Cloth, limp, 6d.

A FIRST LATIN COURSE. By G. B. GARDINER, M.A., D.Sc., and A. GARDINER, M.A. viii+227 pages. 2s.

A SECOND LATIN READER. With Notes and Vocabulary. By GEORGE B. GARDINER, M.A., D.Sc., and ANDREW GARDINER, M.A. Cloth, 1s. 6d.

A LATIN ANTHOLOGY FOR BEGINNERS. By GEORGE B. GARDINER, M.A., D.Sc., and A. GARDINER, M.A. Cloth, 2s

A LATIN TRANSLATION PRIMER. With Grammatical Hints, Exercises and Vocabulary. By G. B. GARDINER and A. GARDINER, M.A. 1s.

LONDON : EDWARD ARNOLD, 41 & 43 MADDOX STREET, W.